Plastic Surgery

The Ultimate Introductory Guide to Cosmetic Surgery And What You Must Know About It

Table Of Contents

Introduction

First off, I really want to thank you for downloading this book. The pages in this short e-book were developed through years of experiences that I have gone through, as well as what has proven to work for others that I have talked to and researched. I also want to congratulate you for taking the time to understand cosmetic surgery and possibly leading a healthier lifestyle.

This book discusses the advantages and side effects of different forms of plastic surgery. Hopefully after understanding these options, you will have a better idea of whether or not you want to undergo surgery. The benefits outlined here are taken from years of collective experiences and testimonials backed by scientific research.

I can guarantee that you will find this book useful if you make sure to implement what you learn in the following pages. The important thing is that you IMPLEMENT what you learn. A change in physical appearance will not solve all of your problems but it can definitely assist you and allow you to feel better about a physical insecurity that you may currently have. What I am giving you is the information you need to get started and the guidelines you will need to make that journey.

I recommend that you take notes while you are reading this book. This will ensure that you get the most out of the information in here. I want you to feel that you made a purchase that is worth your money and so that you can look over the notes of this book even after you've finished reading it. The notes will help you to pinpoint exactly what you need to implement and by writing things down, you will be able to recall specifics and how to handle certain situations when they arise.

Lastly, remember that everything in this book has been compiled through research, my own experiences, as well as the experiences of others, so feel free to question what you have read in this book. I encourage you to do your own research on the things that you want to look deeper into. There are many myths created by the commercial cosmetic industry, mainly because there is profit to be made off of ignorant consumers. You must be aware of what is true and false and that is one of the reasons why I created this book.

The more you understand about your own health and body, the better off you'll be. Cosmetic plastic surgery will take some preparation, understanding, and planning on your part, but it is definitely a realistic possibility for you!

Chapter 1:

Plastic Surgery by Definition

Plastic surgery is the line of work in the medical field concerned with correcting or restoring form and function of a body part. The term "plastic surgery" is derived from the Greek word "plastikos", meaning "fit for molding", and was coined by Pierre Desault in 1798. Desault used the term "plastic surgery" to describe procedures that repaired facial deformities.

There are two distinct categories in modern plastic surgery: reconstructive surgery and cosmetic surgery. This book will mainly focus on cosmetic surgery.

Reconstructive Surgery

Reconstructive plastic surgery is done to correct skin impairments caused by chemicals, burns, injuries, congenital and developmental abnormalities, infections, diseases, and illnesses. Reconstructive surgery is usually performed to restore or improve function, as well as to restore normal appearance. Examples of reconstructive surgery include hand surgery, laceration and scar repair, tumor removal, cleft lip and palate surgery, contracture surgery, and breast reconstruction.

Plastic surgeons usually use tissues in covering defects. Free flaps of muscle, skin, fat, and bone, are removed from the body, transferred to the desired body part, and reconnected to the blood supply by suturing the blood vessels. However, micro-surgery is used if there is no local tissue available.

Cosmetic Surgery

Cosmetic surgery is an optional surgery in which normal body parts are surgically modified in order to improve an individual's physical appearance. This is common among celebrities who need to look their best on camera and get paid based off their appearance. Another common demographic is older adults who want to manage the signs of their aging by trying to minimize them or make them disappear completely.

Examples of common cosmetic surgeries are tummy tuck, eyelid surgery, breast implants, breast reduction, breast lift, butt implants, nose job, ear surgery or ear pinning, face lift, brow lift, cheek lift, chin implant, cheek implants, and liposuction. To briefly explain some of these procedures, tummy tuck is the trimming and reshaping of the abdomen, brow lift elevates the eyebrows, and forehead lift smoothens the skin of the forehead. Liposuction is a procedure that uses ultrasonic energy, or a traditional suction technique, to facilitate fat removal.

Chronology of Plastic Surgery

There is a long and complex history in the field of plastic surgery. Below are interesting facts regarding the history of cosmetic surgery:

Around 3,000 to 2,500 BC, plastic repair for a broken nose was first mentioned in an ancient Egyptian medical text, dating back to the Old Kingdom. Physicians from Britain visited India in order to see how plastic surgery of the nose, also known as Rhinoplasty, was performed by indigenous means. After the physicians came back home, the Gentleman's Magazine of Britain was able to publish reports about a Kumhar vaidya performing Indian Rhinoplasty in 1,794 BC.

By 800 BC, India was already carrying out reconstructive surgery techniques. Sushruta, an ancient Indian physician, brought significant contributions to the field of cataract and plastic surgery in the 6th century BC.

In 750 AD, Dr. Sushruta's medical works were translated, originally from Sanskrit, to the Arabic language, during the Abbasid Caliphate, making their way into Europe through the use of intermediaries. This allowed the Branca family from Italy to familiarize Sushruta's techniques. The works also provided Gaspare Tagliacozzi, the father of plastic surgery, the thought of transferring skin flaps from the upper arms, to the noses of soldiers who were heavily wounded by swords during battle. By 600 BC, a Hindu surgeon used a piece of cheek in reconstructing a nose.

Because of his 20 years of diligent study regarding local plastic surgery in India, English doctor, Joseph Constantine Carpue, was the first doctor to conduct major operations in the Western world, where instruments written in the Sushruta Samhita, were modified. One of Carpue's successful operations included performing a procedure on a member of the British military, whose nose was severely damaged by the toxic effects of mercury.

The Romans also performed simple cosmetic surgery techniques. They began to repair damaged ears around the 1st century BC. However, due to religious reasons passed down by their Greek predecessors, they did not dissect human beings or animals. Fortunately, Aulus Cornelius Celsus left some accurate records regarding anatomy, which have been of great use in the development of plastic surgery.

In the middle of the 15th century in Europe, Heinrich von Pfolspeundt performed Rhinoplasty by suturing the nose area with skin taken from the back of an arm. This surgery became common in the 19th and 20th centuries. It was also in the 19th century that the creation of anesthesia, disinfectants, and sterile techniques, reduced the pain and infection brought about by surgery. The invention of antibiotics, such as Penicillin and Sulfonamides, made performing Elective Surgery possible.

In 1818, Dr. Carl Ferdinand von Graefe, of Germany, published a piece called Rhinoplastik as his major work. It was he who modified the Italian method of rhinoplasty, using a skin graft from the arm instead of a pedicle flap.

In 1845, Johann F. Dieffenbach wrote a comprehensive text about Rhinoplasty, with the title "Operative Chirurgie", and coined the idea of re-operation, in order to enhance the appearance of a reconstructed nose.

In 1891, an American otorhinolaryngologist by the name of John Roe, experimented in the field of cosmetics by reducing the dorsal nasal hump of a young woman. A year later, Robert Weir attempted to reconstruct sunken noses using xenografts, but was unsuccessful in doing so.

In 1898, German orthopaedic-trained surgeon, Jacques Joseph, published his first record of Reduction Rhinoplasty. In 1928, he published Nasenplastik und Sonstige Gesichtsplastik. In 1923, the first modern procedure of cosmetic Rhinoplasty was done, while the first public face lift was done in 1931.

Magic of Popularity

Plastic surgery was not widely known to the public until around World War I, when it became the pioneering method of repairing the wounds and injuries suffered by soldiers in battle. From that point in time, it gained momentum for the next five decades, but plastic surgery was only limited to repairing injuries. Surgeons were still clueless regarding how to get the appearances of people back to normal and even more clueless regarding how to make them look more aesthetically pleasing; thus, they never associated surgery to cosmetic purposes.

However, times changed during the sixties, when surgical procedures and materials greatly improved. The biggest improvements were better skin graft methods and silicone implant inventions, which could be applied to many different parts of the body. Not long afterwards, plastic surgeons finally realized that they could actually enhance people's physical appearance using these techniques.

It was at this point that the idea of cosmetic surgery was promoted heavily. Over time, more and more "everyday people" started to look for procedures. In the eighties, marketers in the United States started to put even more importance on physical appearance, making plastic surgeries what we call a commonplace.

Not only has plastic surgery grown in popularity but it has also evolved greatly over the last 30 years, mainly due to the fact that techniques are constantly being improved. Surgeries are becoming less invasive and the recovery times have become much shorter. Patients are less threatened when undergoing plastic surgery and the media is giving as much attention to it as ever. As a matter of fact, you could expect a large majority of high profile female figures nowadays to have undergone some type of surgery to enhance her appearance.

Statistics Tell You That...

The British Association of Aesthetic Plastic Surgeons (BAAPS) stated that as of 2013, liposuction was the most in demand procedure, among all, in the United Kingdom, with a rise of 41%. In total, 50,122 procedures were performed, with an average rise of 17% per year, since 2012. The audit has highlighted a notable rise in all cosmetic surgeries - a trend that has never existed since pre-recession.

Breast enlargement has been the most popular procedure, with 11,135 augmentations. Meanwhile, the biggest increase was in Liposuction, with an overall 4,236 Liposuction procedures done during the calendar year of 2012; this was followed by an incredible increase of 41% in 2013. Among women, it ranked sixth place as the most popular, while it was in fourth place among men.

Eyelid operations were second on the list, while neck and face lifts came in third place. Even though none of the top ten surgeries for women decreased in total numbers, some slightly deflated in terms of popularity. From sixth most popular in 2013, Abdominoplasty moved down to seventh place, while fat transfer fell from seventh to eighth place.

For men, Rhinoplasty was the most common cosmetic surgery with 1,037 total surgical procedures carried out in 2012; eyelid operations took second place, followed by Breast Reduction. Also during 2013, Ear Correction surgeries dropped from fourth down to fifth most popular amongst men. In total, men obtained 9.5% of all cosmetic surgeries, with 4,757 procedures, a rise of 16% from 2012.

The results led BAAPS to the conclusion that the public prefers using proven and tested surgical means rather than those magical-sounding quick fixes that have no promising results. Nevertheless, this data does not include Botox injections and other non-surgical "lunchtime" cosmetic methods.

According to the Department of Health in the same year, all UK cosmetic products only amounted up to £2.3bn in 2010 but there is a high chance that the products will rise to an estimation of £3.6bn by 2015. The report also showed that non-surgical procedures comprised of 75% percent of the market value and accounted for nine out of ten procedures.

Pros and Cons

Millions of people worldwide undergo cosmetic surgery each and every year. Plastic surgeons in modern times can change almost any aspect of a person's physical appearance, from head to toe. Even though some cosmetic procedures are conducted to restore optimal health, such as reconstructing a deformed face brought about by an accident, a large percentage of procedures are still voluntary. Weighing the advantages and disadvantages before undergoing any surgery can help a person make the decision whether to push through with the operation or not.

Pros

Enhanced Appearance

The most obvious reason for getting cosmetic surgery is appearance enhancement. For example, a woman with small breasts can get breast implants in order to get that perfectly feminine shape that she has been dreaming of for her entire life.

This can stem from a feeling of insecurity from the woman's youth and some will argue that it will not fix the deeper psychological issues. However, there is also the argument that if it helps a woman to feel sexy and feminine, then it will ultimately lead to her living a happier life.

Increased Confidence and Self-Esteem

Branching off from the first benefit, appearance enhancement can bring increased confidence and self-esteem as a result. Once a person loves the way they look, they usually feel better about themselves and thus, their confidence will increase.

Improved Health

In some cases, cosmetic surgery can actually improve a person's health. To cite an example, a large busted woman can undergo breast reduction surgery in order to alleviate the pain that she is experiencing in her back. On a regular basis, she feels the pain because of the extra pressure pulling on her back and shoulders as a result of her large breasts. When her breast size is reduced, she may feel better about herself and her posture will improve, thereby alleviating the pain.

Cons

Complications

Despite all of the great benefits from getting cosmetic plastic surgery, there is always the opposite side of the coin. Undergoing cosmetic surgery has its own drawbacks, one of which is dealing with the risk of complications.

Manifesting signs of bad reactions to anesthetics and the possibility of bleeding, can be potentially uncomfortable. Other possible complications include unwanted scarring, fluid accumulation, skin loss, deep vein thrombosis, complications with sutures, physique asymmetry, necrosis, blood clots, skin changes, numbness and other sensory disturbances, as well as heart and lung complications.

Unmet Expectations

Not surprisingly, sometimes people expect more than what cosmetic surgery can provide. They have high expectations before the surgery and if the surgery does not fix their problems, they can become jaded. Some people believe that cosmetic surgery will help greatly with their social lives, rather than just getting the surgery for enhancing their appearance. These people often become frustrated if their post-surgery appearance doesn't dramatically change their dating life, for example.

Affordability Issues

It is a part of reality that when one wants to get cosmetic surgery, he or she has to pay a big amount of money. Cosmetic surgery is expensive, and most importantly, it requires time off from work for the operation and recovery. If a person does not have this type of financial and time flexibility, planning the surgery may be difficult for them.

Chapter 2:

Calling The Surgeon

In considering cosmetic surgery, it is essential not to overlook the importance of choosing the right surgeon. The following guidelines can aid an individual in finding the best cosmetic surgeon for his or her needs:

The surgeon should be board-certified and a member of at least one of the organizations listed below. (Note that surgeons from other professional organizations may be qualified, but the groups written below ensure that their doctors are nationally certified to do cosmetic surgery):

The American Academy of Dermatology

The American Society of Plastic Surgeons

The American Academy of Facial Plastic and Reconstructive Surgery

The American Board of Cosmetic Surgery

Among the four organizations mentioned above, the American Board of Cosmetic Surgery is the most certified board and exclusively certifies surgeons to perform cosmetic procedures. For the surgeon to be board-certified by this group, a surgeon must meet the following criteria:

Satisfactory completion of specialty residency training and board certification in either Dermatology, Obstetrics and Gynecology, Oral and Maxillofacial Surgery, Plastic Surgery, General Surgery, Opthamology, Oculoplastic Surgery, and Otolaryngology.

Performance of at least 100 recorded cosmetic surgical cases is followed by the fulfillment of a one-year fellowship program focused on the procedure in a 12 month period before certification.

Because the cosmetic industry is full of competition, there are many surgeons who will try to cut corners in order to lower the cost of surgery and in turn, lower the quality of the operation. If a price seems too good to be true, conduct further research and collect some referrals from previous patients before making the decision to use that particular surgeon. Upon narrowing the list down to one surgeon, consider these key questions:

What is his/her area of expertise?

Has he/she practiced for years and/or conducted the procedure many times?

If he/she will not perform the procedure in a personal office, does he/she use a certified surgical facility containing updated emergency equipment, anesthesia monitoring devices, and board-certified anesthesiologists? Is the facility accredited?

How much will the surgery cost, including the surgeon's fees, anesthesia, operating room, and other charges?

Am I allowed to see photos of other patients before and after surgery? Are computer images available so that they can be viewed together with the surgeon?

Is the surgeon encouraging questions to be asked?

How many of his/her patients have returned for procedural revisions or corrections?

In case of possible second surgeries, what will be your financial responsibility?

After settling on a surgeon and procedure, a second opinion is needed because confidence matters when making the final decision.

Now that you know what to look for, let's take a look at what to AVOID. Cosmetic surgeons like these should be excluded:

The surgeon does not regularly perform the considered procedure.

The surgeon does not openly discuss the risks of surgery and possible complications.

The surgeon guarantees results and urges you to make your decision at a pace quicker than you are comfortable.

The surgeon does not, will not, and/or can not show other patients' before and after photos.

The surgeon proposes bargaining of fees or gimmicks and does not give you instructions on post-operative care before undergoing the procedure.

Also, it is important to keep in mind that caring for patients does not end after the surgery. The surgeon's policies should involve necessary surgical revisions in case of follow-up procedures. He/she must be willing to receive follow-up visits as well.

Know the Candidates

The American Society of Plastic Surgeons (ASPS) has declared that there are two categories of good candidates for cosmetic surgery. They are as follows:

Patients with a strong self-image and are worried about certain physical features, which they would like to enhance.

Patients with low self-esteem who have a cosmetic or physical defect that is greatly affecting their everyday life.

On the contrary, listed below are the inappropriate candidates, according to the ASPS:

Patients in crisis

Patients who are going through hardships such as the death of a spouse, loss of a job, or divorce, may want to achieve goals that can not be met, like changing their appearance for immediate gratification. These people are in a very vulnerable position and may regret their decision, as it will most likely be made in a reactionary way.

Instead, resolving the crisis itself must be the priority for people in these situations. They must realize that there is no urgency in getting cosmetic surgery done. If it does not affect your health, taking the time to weigh out the pros and cons will help prevent making big decisions under stress.

Patients who have unrealistic expectations

Examples of these patients are those who want to live a celebrity lifestyle by insisting on getting a celebrity's nose, patients who want to get their appearance perfectly back to normal after a serious illness or severe accident, and patients who want to bring back their glorious youth from the past, with no regards for the natural aging process.

Patients who are impossible to please

These patients often consult with a great deal (sometimes dozens) of surgeons since they want to hear the answers they are expecting. They hope for a cure to a certain problem which may not even be physically possible. These patients are especially prone to falling for sly surgeons who can promise them unrealistic results for an amazing rate.

Patients who are perfectionists, even if they only have minor defects

These people believe that their lives will be perfect once their perceived flaws are fixed. However, perfectionists may still undergo the surgery, but only after understanding the reality that results may not meet their initial goals.

Patients with mental illnesses

If a person with a mental illness has goals for surgery that are not related to their psychiatric illness, they can still be appropriate candidates, provided that the surgeon collaborates with their psychiatrist and family.

Nice to Know: Body Dysmorphic Disorder

Researchers believe that an obsession with plastic surgery can be associated to psychological disorders. Body Dysmorphic Disorder (BDD) is imminent with those who are obsessed with getting plastic surgery.

BDD refers to a disorder in which the victim becomes preoccupied with the thought that they have defects on their face or body. While 2% of the population in the United States suffers from BDD, 15% of clients who see a cosmetic surgeon and dermatologist suffer from it.

Half of these sufferers who have cosmetic surgery done are not satisfied with the aesthetic outcome, leading some of them to go so far as to commit suicide. While many of these patients seek out the surgery, the procedure does not actually treat the disorder. Instead, it could worsen the problem. The psychological cause of this disorder is idiopathic, which is why treating BDD is so difficult.

Others say that the obsession with correcting a body part could just be a sub-disorder, like Anorexia Nervosa. In some instances, people who ask surgeons to let them undergo further surgeries, despite the surgeons prohibiting them to do so, still try to perform the surgery themselves. They inject and cut themselves and as an outcome, compromise their own health and safety.

Chapter 3:

What Happens After Surgery?

For most patients, an increase in confidence and self-esteem as the result of a successful cosmetic operation produced a snowball effect in other areas of their lives. Some became less anxious or self-conscious upon socializing with other people. Some became more outgoing since they felt that they were already spared from the scorn of others. Getting cosmetic surgery legitimately was a positive life-changing event for these individuals.

There is no doubt that cosmetic surgery can turn almost anyone who chooses to get it, into a more attractive person. Countless studies have shown that attractive people are generally viewed as more intelligent, more capable, more honest, and more successful. Attractiveness is highly attributed to variations in salary levels, hiring decisions, and professional recognition and promotions. Fair or not, we must accept reality.

Some studies have also shown that when attractive people are in restaurants and retail establishments, they receive better and quicker service. There are also increased romantic opportunities, as well as opportunities to meet people in a cold-approach setting. This suggests that improvement in a person's physical appearance may be treated as a blessing to his/her career, status, and social life.

Moreover, cosmetic surgery can have many benefits to a person's health and quality of life. Breast reduction, particularly among women, can cause relief from daily pain, bringing forth a dramatic increase in quality of life. For a patient who has loose, hanging skin weighing 20 pounds, he/she could feel more comfortable upon removal of the skin. His/her level of physical activity would surely increase, positively affecting both his/her physical and mental health. For a disfigured patient, the surgery can make the person feel free and accepted again. The world will no longer stare at them and avoid eye contact - a priceless improvement in mental health.

On the other hand, cosmetic surgery may also give undesirable consequences to an individual's daily life. Depending on the circle of friends that a person has, he/she could become the feed of gossip, due to the jealousy and insecurity of others. The person could also become even more uncomfortable with the way people look at him/her due to self-consciousness regarding the enhanced body part. The possible adverse social and psychological effects of cosmetic surgery must also be taken into consideration before getting the surgery, for the reason that it does not actually solve every problem one has in the long-term.

Time For A Change

With a change in appearance comes a change in diet and lifestyle as well. Cosmetic surgery should not be seen as the only way of improving a person's appearance. The guidelines below are regarded as significant in complementing desirable effects of cosmetic surgery post-operatively:

Consuming light, soft foods is recommended after undergoing Rhinoplasty, as well as with some of the other common cosmetic surgeries. On the first post-operative day, eating of pudding, Jell-O and soups is considerably encouraged.

Staying away from spicy foods and consuming foods that are cool instead, is also advised. Spicy foods and hot liquids will only lead to dilation of the blood vessels, making them more susceptible to bruising and swelling. Elevating the head can be very beneficial during the recovery process as well. Nasal irrigation cleans the nose and icing keeps things cool and less swollen.

Smoking cigarettes, drinking alcohol, and consuming foods high in sugar, caffeine, or salt, should be avoided.

Food and drinks that are hot may burn the mouth because of transient numbness, while those that are cold may prompt coughing.

Walking is best done if there is no over-exertion occurring, which can elevate heart rate and cause bleeding during the first two weeks after surgery. Having assistance when walking is recommended.

After two weeks have passed, a stair master, stationary bicycle, elliptical trainer, and other non-impact exercises can finally be integrated. It is also permitted to add light weight training. Sexual activity is safe to resume at two weeks after facial plastic surgery.

A month after the procedure, running and full exertion is not restricted anymore. At six weeks, return to contact sports is fine. However, wearing swimming goggles is prevented for two weeks following an Eyelid Surgery or a Face Lift, and a month after Rhinoplasty.

Are There Any Alternatives?

Cosmetic surgery is something not everyone can afford because of the funds needed. In this section, alternative cosmetic-like procedures will be discussed:

Botulinum Toxin

Botulinum toxin is a neurotoxin produced by *Clostridium botulinum*, a bacterium. In 2002, injectable botulinum toxin types A and B became FDA approved in their use for improving moderate to severe frown line appearances between the eyebrows.

Injections block signals in the muscular nerve and weaken the muscle. As a result, they are able to diminish facial wrinkles. Results may occur within a few days, but the full effects take up to a week to see. The American Society of Plastic Surgeons said that the typical improvements last for three to four months.

The ASPS reminds everyone who undergoes a Botox procedure that it is best not to rub or massage the treated areas post-treatment because if done so, botulinum toxin can migrate to other areas of the face, resulting in temporary facial weakness and/or drooping. Some of the side effects of Botox include redness, headaches, and nausea.

Dermal Fillers

These injectable products restore a youthful glow to the face by increasing lip plumpness, enhancing shallow contours, and softening creases and wrinkles. Dermal fillers only produce temporary results, so they must be repeated and maintained.

Dermal fillers have two distinct categories. The first category is temporary, while the second category is the semi-permanent route. Human fat, Collagen, Polylactic Acid, Hyaluronic Acid, and Calcium Hydroxylapatite all belong to the first category. The semi-permanent category contains Polymethylmethacrylate alone.

The ASPS warns that the duration of the "over-filled" appearance after treatment depends on a case to case basis. It may resolve within a few hours or a few days, while for others, it could last up to a few weeks. Moreover, fillers made from non-human resources need to undergo a pre-treatment allergy test. Complications from fillers include acne, bleeding, bruising, skin rash, and temporary facial paralysis.

Glycation Reduction

Glycation is the bonding of a lipid, or protein molecule, with a sugar molecule. This process, if continued, forms Advanced Glycation End products (AGEs), which attach to the cells. AGEs can cause complications in people with diabetes, such as retinopathy and nephropathy. In addition, glycation results in wrinkling and other aging signs, by cross-linking the collagen fibers.

Glycation is also reduced by taking glycation-inhibiting supplements, foods, and substances, combined with a low glycemic index diet. Glycation-inhibiting supplements and substances include *Panax ginseng* extract, Carnosine (amino acid), guava, and Benfotiamine.

Oxidation and protection against UV Exposure

Photo-aging exacerbates skin aging. The UV effect on the skin involves the generation of Reactive Oxygen Species in the cellular matrix of the body, leading to additional deleterious effects when it comes to skin aging.

The use of antioxidants has been proven to reduce fine lines and wrinkles, as well as improve the moisture and elasticity of the skin. Beta-carotene improves elasticity and facial wrinkles. Lycopene protects the skin from potential photo-damage.

Replenishment of Hyaluronic Acid

This Nonsulfated Glycosaminoglycan is widely used in the cosmetic industry because of its potentially strong water-binding property. One study has shown that it can improve skin hydration and overall elasticity.

In one study, a combination of Hyaluronic Acid and Retinol-derivative, Retinaldehyde, showed significant improvements in overall photo-aging, perioral wrinkles, and nasolabial folds. Clinical photo-aging signs on the face and hyperpigmentation also improved.

Circulation Enhancement

According to some research, a reduction of blood flow is one of the main culprits in having wrinkles among women. Omega-3 Fatty Acid supplements help to promote micro-circulation, reduce facial aging, and improve endothelial function. In addition to that, Omega-3 Fatty Acids can also protect the skin from UV injury. Application of topical Eicosapentaenoic acid can reduce epidermal thickening and inhibit collagen decrease, as induced by UV light.

Conclusion

I worked hard on creating the best guide for plastic surgery that I could. Hopefully it helped you with deciding on whether or not you'd like to pursue the idea of surgery to aid in your physical appearance. Maybe it also motivated you to start shedding those extra pounds without the help of surgery.

Regardless, if you are interested, check with your doctor and family before you get started and don't be afraid to share this book with someone who might be able to use it.

If you feel like you learned something from this book, please take the time to share your thoughts with me by sending me a message. I would also appreciate it if you left a review on Amazon! I try to respond to as many messages as I can!

Thank you and good luck on your journey!